STRIPES
AN AIRMAN'S BULLET WRITING (AND CAREER MANAGEMENT) GUIDE

HAMP LEE III

MONTGOMERY, ALABAMA

Copyright © 2017, 2019 by Hamp Lee III.

All rights reserved. No part of this publication may be reproduced, distributed or transmitted in any form or by any means, including photocopying, recording, other electronic or mechanical methods, or information storage and retrieval systems, without the prior written permission of the publisher, except in the case of brief quotations embodied in critical reviews and certain other noncommercial uses permitted by copyright law. For permission requests, write to the publisher at info@commissionpubs.com.

Recommendations within *Stripes: An Airman's Bullet Writing (and Career Management Guide)* are for informational and educational purposes only. Please consult with the appropriate (and respective) professionals, agencies, or groups before acting on the information in this book.

The information in this book does not represent an endorsement by the Department of Defense or United States Air Force. Views are entirely representative of Hamp Lee III.

Stripes: An Airman's Bullet Writing (and Career Management) Guide
Hamp Lee III — 1st ed.

Library of Congress Control Number: 2017904697
ISBN: 978-1-940042-52-7

CONTENTS

INTRODUCTION	5
1. ENLISTED EVALUATION SYSTEM	7
2. CAREER MAP	11
3. WORD PICTURE	23
4. BULLET ELEMENTS	27
5. DUTY DESCRIPTION	49
6. AF FORM 910 LINE-BY-LINE	51
7. AF FORM 911 LINE-BY-LINE	59
8. PROMOTION BOARD SCORING	67
CONCLUSION	71
ACTION VERB LIST	73

INTRODUCTION

The Enlisted Performance Report (EPR) is one of the most important documents in an enlisted member's career. This form has long-standing impacts during and after his or her military service. The final rank on the DD Form 214, *Discharge Papers and Separation Documents*, is mostly based on the information provided in his or her evaluations.

Evaluations help make important personnel decisions, such as promotion consideration, career opportunities, retraining, and reenlistments. With the significant weight EPRs have in your career and beyond, I wanted to write a guide to help you map out your career expectations and goals, outline specific bullet writing mechanics, and write great EPRs line-by-line.

Stripes: An Airman's Bullet Writing (and Career Management) Guide is the culmination of over a decade of teaching on bullet writing. Since starting this journey in 2008 as a Technical Sergeant (TSgt), I continued to be a career-long learner, tweaking and adjusting my briefings based on what I learned along the way. And this book is no exception.

This book has been updated to provide several important updates since its original release. I have expanded several discussions and topics throughout the book and added a chapter about promotion board scoring.

Our Airmen today have so much greatness and potential in them. The stripes they wear do not give justice to the abilities, leadership, and intellect that resides within them. Our Airmen need supervisors and mentors who will make deliberate investments in their lives and careers; to help them unlock the potential that exists within them so they can reach their professional and personal goals and serve their nation with honor and distinction.

I stand with many other supervisors, mentors, and leaders who invest in Airmen every day. And I hope *Stripes: An Airman's Bullet Writing (and Career Management) Guide* will help you document your duty performance, meet your expectations and goals, and reach your Point B.

O N E
ENLISTED EVALUATION SYSTEM

Air Force Instruction (AFI) 36-2406, *Officer and Enlisted Evaluation Systems* lists three purposes for our enlisted evaluation system:[1]

1. establishes performance standards, meaningful feedback, and direction on how to meet established standards and expectations,

2. provides a reliable, long-term, and cumulative record of performance and promotion potential based on that performance,

3. gives sound information in identifying the best qualified enlisted personnel for promotion and other personnel management decisions.

Because the enlisted evaluation system focuses on performance, a supervisor's role is very important.

---

[1] Air Force Instruction 36-2406, Officer and Enlisted Evaluation Systems. Paragraph 1.1. United States Air Force. 8 November 2016. http://static.e-publishing.af.mil/production/1/af_a1/publication/afi36-2406/afi36-2406.pdf.

Supervisors must have an adequate understanding of the enlisted evaluation system to ensure Airmen are accurately assessed and rated through their performance in four specific areas:[2]

1. how well the Airman accomplishes his or her job,

2. the quality the Airman brings to the job,

3. development of skills and leadership abilities,

4. determining who will be selected for advancement (e.g., promotions, assignments, and other personnel actions).

Though a supervisor does not select his or her Airman for advancement, his or her assessment of the Airman is a part of the record for promotion selection or other career opportunities. Through the use of performance feedbacks and evaluations, supervisors can appropriately (and accurately) identify an Airman's performance.[3] Three resources supervisors can use to establish expectations and identify an Airman's performance are Air Force Handbook (AFH) 36-2618, *The Enlisted Force Structure*, Airman Comprehensive Assessments (ACA), and EPRs.

---

[2] Ibid., 1.1.2.

[3] Ibid.

Enlisted Force Structure

AFH 36-2618, *The Enlisted Force Structure* defines the leadership levels, enlisted force structure tiers (Junior Enlisted Airman, Noncommissioned Officer (NCO), and Senior Noncommissioned Officer (SNCO)), ranks, roles, terms of address, general responsibilities, duty titles, and special positions for enlisted members.[4] It provides the general framework for the enlisted force structure that best meets mission requirements while developing institutional and occupational competencies.[5]

Consider AFH 36-2618 as the blueprint for an Airman's career. Regardless of your rank or position, AFH 36-2618 provides foundational guidance and instruction for Airman expectations and responsibilities.

Airman Comprehensive Assessment

AFI 36-2406, *Officer and Enlisted Evaluation Systems* describes the ACA as the formal communication between a rater and ratee.[6] It communicates responsibility, accountability, Air Force culture, an Airman's critical role in support of the mission, individual readiness, and

---

[4] Air Force Handbook 36-2618, The Enlisted Force Structure. Paragraph 1.2.2. United States Air Force. 5 July 2018. http://static.e-publishing.af.mil/production/1/af_a1/publication/afh36-2618/afh36-2618.pdf.

[5] Ibid., Paragraph 1.2.1.

[6] Air Force Instruction 36-2406, Officer and Enlisted Evaluation Systems. Paragraph 2.1. United States Air Force. 8 November 2016. http://static.e-publishing.af.mil/production/1/af_a1/publication/afi36-2406/afi36-2406.pdf.

performance feedback.[7] ACAs outline expectations on duty performance and how well an Airman is meeting those expectations, to include information to assist the Airman in achieving success.[8] It provides an opportunity to discuss personal and professional goals, as well as assessing or reviewing the objectives, standards, behavior, and performance with the Airman.[9]

Enlisted Performance Report

The EPR is used to document potential and performance as well as provide information for making a promotion recommendation or selection, assignment consideration, school nomination and selection, and other career management decisions.[10] The EPR is also a part of the End-of-Reporting Period Feedback.[11] This feedback provides a review of the previous reporting period and establishes expectations for the new reporting period.

---

[7] Ibid.

[8] Ibid.

[9] Ibid.

[10] Ibid., 1.2.4.

[11] Ibid., 4.18.1.

TWO
CAREER MAP

When traveling to a new location, you often use your GPS or a travel website for directions. You input your current location as well as your desired destination. The results might provide several routes you can take that vary in time, distance, and cost. Based on these factors, you select the most advantageous route to take. Having a career map is no different.

A career map is a professional and personal resource that outlines specific expectations and goals that span your military career. Just like a GPS, each expectation and goal move you closer to your Point B.

Point B

Your Point B represents the final day you wear your service dress uniform—whether through separation or retirement. Consider your service dress uniform and the ribbons, opportunities, awards, and promotions you received along the way.

Hopefully, the career map on the following pages will be a great resource in helping you accomplish every expectation and goal toward your Point B. I will use the rest of the chapter to outline a career map scenario between a supervisor and subordinate, Sergeant Jones and Airman Snuffy.

Jones & Snuffy

After their initial feedback, Sergeant Jones calls Airman Snuffy into his office. He tells Airman Snuffy he wants to outline a career map for him. Sergeant Jones tells him a career map is one way he can set expectations and goals throughout his career, like driving from Point A to Point B. But for now, they will create a career map for the next three years.

Because Airman Snuffy is relatively new to the Air Force, a lot of the information for his career map will come from the ACA they recently completed. Though a career map might have some of the same information as an ACA, it can be a personal resource that is not shared in official or professional settings.

Sergeant Jones tells Airman Snuffy there will be several steps to consider for his career map. For most steps, he will only identify three to five items, but some steps might require more.

Big Rocks

Big rocks represent three to five significant goals you would like to achieve while assigned at your current base and rank/enlisted force structure tier. These 'big rocks' will often be significant accomplishments in your career.

Sergeant Jones believes Airman Snuffy is off to a great start in his career. After reviewing his ACA and speaking with him, they establish the following three 'big rocks' while assigned to the unit and base:

1. Be selected as a Senior Airman (SrA) Below-the-Zone.

2. Be selected for promotion to Staff Sergeant (SSgt) on his first attempt.

3. Earn his Community College of the Air Force (CCAF) degree.

With these 'big rocks,' Sergeant Jones knows Airman Snuffy will need to accomplish many things along the way. Airman Snuffy cannot simply walk into a testing room, Junior Enlisted Airman Council (JEAC) meeting, or Central Base Board. He will need to put in some work—in his unit, across the base, and in the local community. Following are the many things Sergeant Jones will need to consider and forecast for Airman Snuffy to have an opportunity to meet his 'big rock' goals.

Small Rocks

Smaller rocks represent the items (or steps) you might need to accomplish to meet your 'big rock' goals.

Professional Military Education (PME) and Training Requirements. Identify the mandatory PME and training requirements for your rank or enlisted force structure tier.

Complete your PME and training requirements as soon as you are able unless extenuating circumstances exist.

To outline Airman Snuffy's PME and training requirements, Sergeant Jones considers the mandatory education and training requirements he needs to complete within the next 12-18 months:

a. First Term Airman Center (FTAC) attendance.

b. Complete Career Development Course (CDC).

c. Complete all 5-skill level upgrade training (UGT) requirements, as identified in his Career Field Education and Training Plan (CFETP).

Work Center Training Requirements and Certifications. Identify the training and certifications required to fulfill your assigned duties. Complete these requirements as soon as you are able unless extenuating circumstances exist.

Airman Snuffy's UGT will satisfy his training requirements. No certifications are required.

Leadership. The Airman Handbook defines leadership as "the art and science of accomplishing the Air Force mission by motivating, influencing, and directing Airmen."[12] Regardless of rank, every Airman can assume leadership

---

[12] The Airman Handbook. Paragraph 10.2. United States Air Force. 1 October 2017. https://static.e-publishing.af.mil/production/1/af_a1/publication/afhandbook1/afhandbook1.pdf.

roles in his or her unit, across the base, and within his or her functional or local community.

Breadth of Knowledge and Experience. Breadth of knowledge and expertise comes through successful performance of varied, yet increasing responsibilities, special duties, and rotating duty sections and assignments.

With each opportunity, provide strong duty performance: leading, problem-solving, and process and product improvement. Demonstrate strong competence, initiative, dependability, and influence.

Mentoring. The Airman Handbook defines mentoring as "a relationship in which a person with greater experience and wisdom guides another person to develop both personally and professionally."[13] Like leading, every Airman can mentor others. He or she merely shares his or her experience and information.

As Airman Snuffy has learned the 'fine art' of cooking in the dorm dayroom microwave, Sergeant Jones helped him organize a cooking class. With the help of the dorm manager and dorm council, Airman Snuffy provided 50 dorm residents with several recipes for cooking healthy meals in a microwave.

---

[13] The Airman Handbook. Paragraph 10.18. United States Air Force. 1 October 2017. https://static.e-publishing.af.mil/production/1/af_a1/publication/afhandbook1/afhandbook1.pdf.

Professional and Personal Education. Identify professional and personal education opportunities and/or goals. For many Airmen, this will first begin with completing their CDCs before moving to other postsecondary education or civilian certifications.

Because Airman Snuffy will spend the next few months completing his CDCs, Sergeant Jones will set a goal of completing one college class within the next year. Airman Snuffy's career map will increase the number of projected classes he should take to meet his 'big rock' education goal.

Unit, Base, and Community Involvement. Show significant impacts in your unit, across the base, and in your local community. You should not participate in every available volunteering event. You should be very deliberate about your volunteering efforts from year to year.

For example, one year you might support the unit booster club and mentor at FTAC. Toward the end of the year, you lead a private organization event. Then you are voted to serve on next year's private organization executive council. With this new executive position, you then taper what you are doing in the booster club, but you continue to volunteer and support their events throughout the year.

Awards. Be deliberate about award submissions. Do not focus on winning a quarterly award several times in a row. Show variety in your award nominations and wins (e.g., unit, wing, Numbered Air Force, Major Command (MAJCOM), functional, etc.). If necessary, build an award submission schedule. Be sure to speak with your supervisor/subordinate about award submissions and the importance of being recognized for superior performance, when warranted.

Sergeant Jones is hoping Airman Snuffy will perform well enough to compete for a base Honor Guard quarterly award within the next 18 months. But he will also be on the lookout for other award opportunities within the unit and their functional community.

Review. Periodically review your career map to ensure you remain on track for meeting your expectations and goals. Add and subtract expectations and goals as necessary. Allow your career map to be a living document that supports and enhances your progression in the Air Force.

Though I do not recommend reviewing your career map every day, my suggestion is to review your career map:

a. after each evaluation period,

b. at the start of each calendar year,

c. before arriving at a new duty section or base.

In addition to these three reviews, I also recommend reviewing your previous evaluations being considered for the Enlisted Force Distribution Panel and promotion board (as applicable) and AFH 36-2618. The career map you build today is based on your previous achievements. Your career map review should be the foundation and catalyst for your next expectation and goal.

If your evaluations have similar accomplishments from previous years, look to show a greater breadth of knowledge and experience on the next evaluation. Speak to your supervisor and leadership chain (as appropriate) for additional opportunities. Review the next higher grade and enlisted force structure tier in AFH 36-2618 to shape a deliberate word picture of your performance. You want to show you are ready for the next opportunity in your military career.

Feedback. As you build your career map, speak to your supervisor, mentors, leaders, family members, and others you trust. Share your thoughts about your desired expectations and goals while at a particular rank or base, or within a specific timeframe. Their perspectives can help you further shape your career map. Welcome their honest feedback and sage counsel.

Having a successful career is not about being a Chief Master Sergeant (CMSgt) or remaining in the Air Force for a specific amount of time. This is your career. Be proud of your service and how you will wear your uniform—with pride, humility, and thankfulness.

Summary of Airman Snuffy's Goals

Big Rocks

1. Be selected as a SrA Below-the-Zone.

2. Be selected for promotion to SSgt on his first attempt.

3. Earn his CCAF degree.

Small Rocks

Year 1

a. CDC completion in three months with a score of ninety percent or above on the end-of-course CDC test.

b. Complete one college course.

c. Serve as a significant supporter in the unit Booster Club and JEAC.

d. Serve as a base Honor Guard member.

e. Volunteer for one base-wide or local community event.

Year 2

a. Lead one large section project (typical SSgt assignment).

b. Complete at least two college courses.

c. Serve on the Booster Club or JEAC executive council.

d. Volunteer for three local community events.

e. Lead one local community event.

f. Earn one quarterly and one annual award (unit, base, or functional).

Year 3

a. Lead two or three large section projects (typical SSgt assignments).

b. Complete at least three college courses.

c. Serve on a different executive council on base or in the local community.

d. Volunteer for one base-wide event.

e. Lead one base-wide event.

f. Provide one speech on base or in the local community.

g. Earn two quarterly and one annual award (unit, base, or functional).

h. Receive a Must Promote or Promote Now on first eligible evaluation.

Big Rock: Earn SrA Below-the-Zone.

THREE
WORD PICTURE

Former Chief Master Sergeant of the Air Force James Cody once explained that the Air Force is not looking at numbers to determine an Airman's performance, but a word picture that describes his or her performance..."you either perform to a certain level or you do not, and we want to be able to describe that in words."[14]

A word picture establishes your performance history and differentiates your performance among your peers. It shows how you might be the best qualified for promotion and other career opportunities. Illustrating the right word picture requires deliberate development on the part of every supervisor and Airman.

Whether you realize it or not, you have had a word picture since you entered the Air Force. When other Airmen consider you, they describe your performance with a specific word picture based on your conduct and performance.

So how do Airmen build a strong word picture?

---

[14] "CMSAF Cody shares advice on EPR, AF changes ahead." Scott Air Force Base. Accessed 7 March 2017. http://www.scott.af.mil/News/Article-Display/Article/766887/cmsaf-cody-shares-advice-on-epr-af-changes-ahead.

Competence + Initiative + Dependability = Influence

1. Competence. Competence is the ability to do something successfully or efficiently. It encompasses the education and training that demonstrates skills, abilities, and knowledge. Competent Airmen know their job and perform it very well.

2. Initiative. Initiative is the ability to assess and initiate things independently. With initiative comes originality, self-motivation, resourcefulness, creativity, inventiveness, imagination, and ingenuity. Competence Airmen show initiative by discovering ways to improve their daily tasks, processes, and procedures. And through each act of initiative, an Airman's sphere of dependability and influence widens.

3. Dependability. Dependability is the quality of being relied upon and counted on. When Airmen display competence and initiative, supervisors increase their trust and confidence in them. They know their Airmen can (and will) get the job done.

4. Influence. If you show competence, initiative, and dependability, your supervisor (and other leaders) will provide you with increasing opportunities to lead and excel. You have proven to be trustworthy and capable of completing detailed and challenging tasks—which expand and enhance your leadership and duty-related abilities. This provides you with greater opportunities to influence Airmen in your unit, on your base, within the local community, and beyond.

The influence you gain helps you build a strong word picture. The word picture you create shapes your evaluations and helps you accomplish your expectations and goals. Having a strong word picture is good, but you must be able to describe your performance through your evaluations.

FOUR
BULLET ELEMENTS

Bullet statements are the building blocks for award statements and evaluations. If you cannot accurately describe your word picture in a bullet statement, reviewers might not understand the impact of your performance and your ability to serve in the next higher grade or career opportunity.

Jones & Snuffy

Sergeant Jones saw great work performance from Airman Snuffy. He was exceeding Sergeant Jones' expectations. Airman Snuffy was also surpassing his career map expectations and goals.

At the beginning of the new quarter, Sergeant Jones brings Airman Snuffy into his office. Sergeant Jones tells him how well he is performing and that he wants to consider him for a quarterly award in a few months.

To find additional opportunities for Airman Snuffy, Sergeant Jones speaks to his 5/6 Council, first sergeant, and squadron superintendent. He also assigns Airman Snuffy other self-improvement courses and leadership opportunities in the unit and around the base.

1. Get the details. Collect the great things you have done for the quarter, reporting period, etc. There are many ways you can keep track of your accomplishments:

a. Weekly Activity Report (WAR). A WAR is a written statement of accomplishments during a given period. Most WARs are weekly submissions to a supervisor. The accomplishments may or may not be written in a bullet format. The format depends on the requestor's requirements.

b. Spreadsheets. Spreadsheets can be used to record your accomplishments by month, quarter, year, etc. Spreadsheets can also be organized in several bullet formats.

c. E-mails. Airmen also use their e-mail Sent Items as a bullet statement repository. These e-mails often describe their coordination and job completion over a specified period. Airmen using this method should maintain current backups of all saved information.

Whatever method you use to document your duty performance, please ensure:

a. you can sustain it for an extended period,

b. it fits your personality or organizational style,

c. it is a medium you are familiar with (paper or electronic).

Jones & Snuffy

With their plan in place, Airman Snuffy heads off into the 'wild blue yonder' to do great things. Each week, he provides a WAR to Sergeant Jones. After reviewing a few of Airman Snuffy's WARs, he makes a few adjustments and adds a few other 'opportunities to excel' to help him along the way.

With the award suspense date approaching, Sergeant Jones reviews the information he collected from Airman Snuffy. He ensured Airman Snuffy accomplished more tasks than bullets he needed. This allows Sergeant Jones to be more deliberate in shaping the word picture for Airman Snuffy's award package.

2. Bullet detail. Each bullet statement should have four parts: accuracy, brevity, specificity, and perspective. The first three are covered in The Airman's Handbook and Tongue and Quill. They are a part of the finishing touches of a bullet statement. But I would like for you to consider these parts earlier in the bullet writing process.

a. Accuracy. Make sure any information you include in a bullet statement is accurate. Using incorrect or false information might show a lack of integrity on your part and cause you to lose the trust of your leadership. With a loss of confidence from your leadership, every document you submit from then on could be questioned and heavily scrutinized.

b. Brevity. Use words in your bullet statements that are simple, exact, and concise. Make each word count.

c. Specificity. Be specific in your descriptions of performance. Examples like '50+' or '40%+' are general statements that can be viewed as a space filler. If a number is more than 50%, say so...53%.

d. Perspective. A perspective displays a particular aspect of performance in one bullet statement. The source information for developing a specific perspective of performance comes from AFH 36-2618, ACAs, CFETP, and EPRs. Each identifies specific instructions for current and future positions, ranks and responsibilities, and enlisted force structure tiers.

Each bullet statement should provide different perspectives that complement one another to form a single word picture. This word picture should 'show' that you are ready for the next promotion or career management opportunity. Therefore, your career map and word picture should align with information identified in those higher positions, ranks and responsibilities, and enlisted force structure tiers.

Jones & Snuffy

Sergeant Jones reviews the information for Airman Snuffy's first bullet statement: I worked with a contractor to buy 13 vehicles for the unit. Everyone said the vehicles helped them do their jobs better.

Though a good start, Sergeant Jones does not believe this is enough information. He will need to know how these new vehicles benefitted the unit.

3. Bullet mechanics. The standard Air Force bullet format, as written in The Airman's Handbook and Tongue and Quill is Accomplishment—Impact (impact and result). Within this chapter, I will go over this format, as well as a few other ways you can structure a bullet statement.

Example:

- Organized weekly 1st Sgt/Right Start briefings; revised slides/organized briefs--prepp'd 2K Airmen...combat ready

—	Bullets start with a single dash and space directly after.
;	Transition between accomplishment and impact.
— —	Transition to the final result of the bullet.
'	Used as an abbreviation, often to save space.
...	Indicates an omission of a word.
/	Often used in place of conjunctions and prepositions.

4. Bullet format. Bullet formats are the foundational elements that shape your bullet statements. They should describe what you did, how you did it, and how it impacted others.

Determining the bullet format you should use will often depend on the information you collect. Following are a few bullet formats to consider:

a. Accomplishment—Impact—Result

- Led unit homeless shelter food drive; collected $8K in food/ 1K lbs clothing--lifted spirits during economic hardships

The accomplishment describes a completed action. It often begins with a strong, complementary action verb. Though it is important to find the 'right' action word or statement, do not get too technical or extravagant in your description. Use the fewest words possible to illustrate your accomplishment accurately. Remember accuracy, brevity, specificity, and perspective.

The impact is the immediate benefit of your accomplishment. It shows how your actions affected, impacted, or benefitted others in your work center, unit, base, Air Force, Department of Defense, etc.[15] Looking at the example, the immediate benefit (or impact) was collecting $8,000 in food and 1,000 pounds of clothes.

The result is the final outcome (or benefit) of the specific accomplishment. The result will often provide quantitative data. I will go over this in greater detail a bit later in this chapter.

---

[15] Air Force Handbook 33-337, Tongue and Quill. Page 251. United States Air Force. 27 July 2016. http://static.e-publishing.af.mil/production/1/saf_cio_a6/publication/afh33-337/afh33-337.pdf.

Bullet Elements 33

In leading the homeless shelter food drive, the result was lifting spirits during economic hardships.

b. What—How—Impact

- First responder to snake bite poison emergency; initiated medical care until Paramedics arrival--saved child's life

As described, the What—How—Impact bullet format states 'what' you accomplished, 'how' you accomplished it, and the impact of your actions. This bullet format can be used when you want to show the impact of how you accomplished a specific task. This is often important when you accomplish a task that is typically assigned to others of a higher rank or position.

c. Lead-in—What—Impact

- Alternate unit fitness program manager; drafted two new squadron policy letters/aided primary--kept Sq fit to fight

A lead-in is simply an introduction. Lead-ins for bullet statements can be an announcement of information, such as 'alternate unit fitness program manager' or 'Hard-charging NCO!' But be mindful with your use of lead-ins in bullet statements. Ensure your use of lead-ins strengthen a bullet statement rather than serve as a space filler.

When deciding on a specific bullet format, there is no right or wrong way to structure your bullet statements. You only need to collect the 'right' information and use the most appropriate bullet format to highlight a specific perspective of your accomplishments.

Jones & Snuffy

Considering Airman Snuffy's first bullet statement, Sergeant Jones believes the Accomplishment—Impact—Result bullet format will be the most beneficial. He thinks he will be able to show what was accomplished, provide an impact, and highlight the result for the unit.

5. Word choice. Word choice is critical in an award statement or evaluation. Do not only look for words that fit on a line. Take the time to find the 'right' words that best describe and complement an accomplishment and/or impact. Again, do not try to be overly technical or extravagant in your description. You want to ensure reviewers can easily understand the significance of your accomplishments.

6. Be consistent in spelling (e.g., CMSgt...Chief or and...&). If you use a particular spelling or abbreviation for one word, try to use the same standard throughout the award statement or evaluation. We will often spell one word several different ways to make it fit on a line. Sometimes CMSgt fits, then in another bullet statement, Chief...or developed on one line, then dev'd in another. But as award statements and evaluations each represent one document, seek to standardize your formatting.

7. Do not vomit all over your report. I have seen evaluations where it seemed like an Airman tried to include everything he or she did in the specified period. EVERYTHING. You do not have to squeeze every possible detail into a bullet statement, award statement, or evaluation. Be very deliberate and intentional. Strive to list no more than two 'connecting' or similar accomplishments in one bullet statement.

8. Every bullet statement is a story. Do you remember writing a book report in school? You could have been assigned to read a 200-page book and write a 5-page paper about the book. In your report, you would select the most pertinent information not only to describe the book but also illustrate your knowledge of it to receive a high grade. Writing a bullet statement is the same way.

9. Minimize acronyms and abbreviations. When using acronyms and abbreviations, ensure your bullet statements remain clear, concise, and understandable. There is no guarantee that only Airmen in your functional community will review your award statements and evaluations. Also, make sure your acronyms and abbreviations are locally approved. Your installation should have a writing guide that provides a listing of approved acronyms and abbreviations.

10. Avoid fluff. "Fluffy" bullet statements take up valuable space you could use to tell your story. Reviewers might not be impressed by such unnecessary words. They might even view your bullet statements negatively.

11. Watch for white space...unless done on purpose. White space and providing minimum bullets can send a message about an Airman's performance (or lack thereof).

12. Connect the dots. As I briefly stated in #7 and 8, ensure the information you provide in a bullet statement tells a story that the reviewer can understand. Providing several accomplishments in one bullet statement and saying 'saved $2M' at the end is not a strong impact when the reviewer cannot understand what you did to save that amount.

Bullets statements with multiple accomplishments often require an appropriate lead-in and/or impact statement to 'connect' them.

13. Quantitative data. Quantitative data expresses a certain quantity, amount, or range when it relates to numerical quantities such as measurements or counts.[16] For enlisted evaluations, providing quantitative data is very important.

Whether there is a description of time or money saved or an increase in productivity, quantitative data can show how you improved a process or product. It describes how your accomplishments benefited (and impacted) your work center, unit, base, functional community, and Air Force.

Quantitative data is often connected with the 'result' of your bullet statements. But depending on the perspective you

---

[16] Quantitative Data. Organisation for Economic Co-operation and Development. Accessed 17 February 2019. https://stats.oecd.org/glossary/detail.asp?ID=2219.

want to share, quantitative data might be better suited in other sections of the bullet statement as well.

Jones & Snuffy

With this first bullet statement, Sergeant Jones works to find the 'right' active verb. Now, he knows verbs must be in the past tense and should be strong and action-oriented. So he goes to his handy verb list (see listing in the back of the book). Sergeant Jones finds 'procured' on his list. Other associated words with procured are acquired, ordered, and budgeted. Though he likes the other options, Sergeant Jones settles on 'procured' to lead-off the bullet.

Sergeant Jones also speaks to the unit Resource Advisor to find out how many vehicles were replaced and how much they cost. The Resource Advisor tells him the vehicles cost $200,000 and Airman Snuffy helped to replace 45% of the vehicles.

Here's how Sergeant Jones structured the accomplishment and impact:

- Procured 13 tactical response vehicles worth $200K; 45% of fleet new/modernized--

Sergeant Jones includes quantitative data in these sections of the bullet statement to highlight the total cost of the vehicles and percentage of vehicles that were replaced. It is well known in their functional community that many junior enlisted Airmen typically do not handle purchases worth

$200,000. Sergeant Jones is providing a deliberate word picture that shows how Airman Snuffy is operating at a level higher than the expectations of his current rank and position.

Sergeant Jones believes this is a great start to describe Airman Snuffy's accomplishment. But he still needs to find the result.

14. Develop the result. As I stated previously, the result is the final outcome of an accomplishment and shows how your actions affected or benefitted others and the level of impact.[17] To find the result for an accomplishment, there are six questions you will need to ask the 'right' person in the process: who, what, where, why, when, and how (as applicable).

Jones & Snuffy

Sergeant Jones remembered Airman Snuffy saying that everyone in the unit said the vehicles helped them do their jobs better. He figured this was a description of an improvement in productivity (time). To find out whether there was an improvement in response time, Sergeant Jones speaks to the desk sergeant. He asks the desk sergeant how many responses did the unit have in the month before the new vehicles arrived and the response time for each. Sergeant Jones also asks the same questions for the month after the new vehicles arrived. With the older vehicles, he

---

[17] Air Force Handbook 33-337, Tongue and Quill. Page 251. United States Air Force. 27 July 2016. http://static.e-publishing.af.mil/production/1/saf_cio_a6/publication/afh33-337/afh33-337.pdf.

learned the unit had 100 responses with a response time of 7 minutes each. For the new vehicles, there were also 100 responses with a response time of 5 minutes each.

Sergeant Jones calculates the response times. 700 minutes for the older vehicles and 500 minutes for the new. He subtracts the two numbers for a 200-minute difference between months. After dividing 200 and 700, he receives the following number: 0.285714285714286. Sergeant Jones uses this figure to describe the percentage of improved response time between the months, which equals a little over 28%. Because the number is in-between 28 and 29, he uses 28% rather than 28.5% because the .5 would be added space.

After a bit of investigating, Sergeant Jones believes he found the final result. He puts together Airman Snuffy's bullet statement in the following manner:

- Procured 13 tactical response vehicles worth $200K; 45% of fleet new/modernized--improved response time by 28%

There are a lot of different ways to describe this bullet. Let's say Airman Snuffy established a new purchasing method that sped payment processing by 55%, cut errors by 25%, and was benchmarked across the MAJCOM. The bullet statement above might not provide the right perspective of Airman Snuffy's performance. Consider the following:

- Fiscal giant! Bought 13 unit vehicles f/$200K; shaped new purchase method--cut errors 25%/ACC benchmarked

I left out 'sped payment processing by 55%,' but if I can include it and ensure the bullet statement is still understandable, I will. But my purpose here is to provide a perspective of his performance that shows leadership, process and product improvement, and a significant impact across a MAJCOM. I also sought an appropriate lead-in, but in this case, it might not be necessary. I could use that space to include the payment processing.

Let's try a few more of Airman Snuffy's bullet statements:

Exercise 1

Airman Snuffy said he was also a Combat Arms Training and Maintenance (CATM) instructor. He said he trained a bunch of Airmen. He made sure everyone was trained before they deployed.

What questions can Sergeant Jones ask the CATM staff or others?

How many Airmen were trained?

How many Airmen deployed?

What timelines or deadlines were met?

What were the specific accomplishments of those trained through the program? How did they use their training?

Are there any metrics from the previous quarter or year?

Is there a way to show improvement or specific target goals being met?

Were there any particular scheduling improvements for the personnel trained?

Sergeant Jones found out Airman Snuffy managed the CATM program. CATM trained 3,600 Airmen, and 1,400 were deployers. One of Sergeant Jones' buddies that went through the training said what his team learned kept their Forward Operating Base (FOB) from being overrun. (This is just an example. Remember Operations Security.)

After thinking about it, Sergeant Jones believes the What—How—Impact bullet format will work best:

- Managed base CATM program; taught 3.6K Airmen/ qualified 1.4K deployers--training kept FOB from being overrun

Exercise 2

Sergeant Jones drafted another bullet statement for Airman Snuffy. He used the Lead-in—What—Impact bullet format:

- Alternate unit fitness program manager; drafted two new squadron policy letters/aided primary--kept Sq fit to fight

How can Sergeant Jones make the impact stronger?

What are some questions he can ask? Who would he ask?

The primary Unit Fitness Program Manager (UFPM) should have the number of failures before and after Airman Snuffy's appointment. The UFPM should also know how many people achieved an 'excellent' score before and after as well. The UFPM informs Sergeant Jones that under Airman Snuffy's new training program, fitness failures decreased by 40% and Airmen receiving 'excellent' scores increased by 3%.

Updated bullet statement:

- Alternate UFPM; drafted CC policy ltrs/est new trng regimen--raised unit 'excellent' Amn 3%/lowered failures 40%

Exercise 3

- Completed 18 semester hours towards Baccalaureate from American Military University--maintained 3.84 GPA

Many Airmen have seen and used a bullet statement similar to this one. Instead of leaving this bullet statement as is, Sergeant Jones asked Airman Snuffy how did he apply the information he learned. Airman Snuffy described several hiring procedures he was able to improve as a result of the classes he took. Sergeant Jones adjusted the bullet with the following information:

- Completed 18 credit hrs; earned 3.8 GPA--applied skills to improve unit hiring procedures from 90 to 10 days

or...

- Earned 18 credit hrs in HR; received 3.84 GPA--applied new skills to improve unit hiring procedures by 80%

Exercise 4

Airman Snuffy received the following bullet statement on his letter of evaluation when Sergeant Jones became his supervisor:

- Tobacco cessation class champion--briefed/promoted CSAF goal of 20% reduction in tobacco use AF wide[18]

As Airman Snuffy's new supervisor, Sergeant Jones believes he can develop a stronger impact for this bullet statement. However, he will need to develop a method for collecting the 'right' information.

Sergeant Jones asks Airman Snuffy to keep a record of the total number of people he trains within his classes. He also speaks to the flight commander. Sergeant Jones asks the flight commander if he can create a survey for Airman Snuffy's tobacco cessation classes. Each quarter, he would like Airman Snuffy to survey each participant to ask if he or she quit smoking, and if so, how much money did he or she save since quitting. The flight commander agrees but would like to ensure each member knows the survey is non-retribution.

---

[18] "4Y AFSC Results." AF EPR Bullets.com. Accessed April 2, 2017. http://www.afeprbullets.com/results.php.

Before the end of the reporting period, Sergeant Jones requests the survey results from Airman Snuffy's tobacco cessation classes. He learns that Airman Snuffy taught every tobacco user on the installation—500 Airmen. Through the training he created, 58% of the participants quit smoking. Combined, the Airmen saved over $5,000 in one year!

Armed with this new information, Sergeant Jones decides the Accomplishment—Impact—Result bullet format is best suited for Airman Snuffy's accomplishment:

- Taught 75 base tobacco cessation classes; created curriculum f/500--reduced smoking 58%/svd Amn $5K in yrly costs

Exercise 5

Sergeant Jones sends Airman Snuffy on a 30-day temporary duty (TDY). He traveled with a team of seven Department of Defense civilians (civilians) to create 1,000 widgets. While TDY, Airman Snuffy created a new automatic oiling function that cut widget production in half.

Sergeant Jones learned it once took 300 hours to produce 1,000 widgets, and now it takes 150 hours. He also finds out from the section chief that this new process is being used across the Air Force.

Sergeant Jones talks to the Resource Advisor and Defense Travel System Administrator to learn the TDY costs for each civilian. Including per diem, each civilian received $5,000 for the 30-day TDY. This is a total of $35,000 for one TDY. Airman Snuffy's new process will save the Air Force $17,500 in TDY costs because the civilians can return after 15 days.

There are several ways Sergeant Jones can format this bullet statement. Following are a few sample bullet statements Sergeant Jones can write depending on the perspective he wants to describe:

- Authored new widget creation process; constructed auto-oiling function--cut production time 50%/saved AF $17K/yr

- Built 1st-ever widget auto-oiling function; cut production 50%/svd 150 man-hrs & $17K yrly--benchmarked AF-wide

- Innovator! Developed new widget production process; created auto-oiling function--saved AF 150 man-hrs/$17K/yr

- Developed new widget creation process; crushed production metric by 50%--saved AF 150 man-hours and $17K/year

15. Watch for errors. Check for spelling and grammatical errors. Even though you might not have written the evaluation, spelling errors on your evaluation become a part of your word picture. Therefore, be sure to thoroughly review your evaluations before signing.

16. Support your claims when asked. If we read all the evaluations across the Air Force in one year, we probably saved over five hundred decillion dollars combined. At some point, you might be questioned for single-handedly saving one billion dollars. So be ready to prove what you write.

Jones & Snuffy

Sergeant Jones selected the following bullet statement for Airman Snuffy's quarterly award package:

- Built 1st-ever widget auto-oiling function; cut production 50%/svd 150 man-hrs & $17K yrly--benchmarked AF-wide

Sergeant Jones thought this bullet statement would be more impactful because it included the impact and result of Airman Snuffy's efforts. When Sergeant Jones' flight chief reviewed Airman Snuffy's award package, she asked to see Sergeant Jones about numbers he provided in the widget bullet statement. While in the flight chief's office, Sergeant Jones outlined the method he used to calculate the data. After hearing his explanation, the flight chief said, "Thank you Sergeant Jones. That will be all."

17. Show variety. Many Airmen often have little to no control over the base and/or work center assignment they receive. They might be required to remain in one position for several years.

As we serve in the places and positions asked of us by our service, I want to challenge you to show variety in each of your evaluations. No two evaluations should list the same information at the same level of accomplishment. This is regardless of assignment or location.

Yes, there are similar things you will accomplish throughout each evaluation period, but you do not have to include them all in your evaluation. Build a comprehensive career map. Speak with your senior leaders. Use information from AFH 36-2618, ACA, and EPR small print to establish a schedule for providing various but deliberate perspectives of performance. Show breadth of knowledge and experience.

If you look back at Airman Snuffy's career map in chapter two, Sergeant Jones was providing an outline for him to complete specific accomplishments at varying levels. Maybe one year he supported the JEAC, but the following year, he would serve as an executive council member.

FIVE
DUTY DESCRIPTION

The duty description is the foundation of the EPR. It should be simple and concise while accurately describing the scope of your key duties, tasks, and responsibilities. Your duty description should be as deliberate as the bullet statements you write. Consider the following process for structuring your duty description:

Line 1—The Lead Off. The first line should outline the entire scope of your duties and area of responsibility. It establishes the beginning of a strong word picture. Consider the word picture for this first line:

Training NCOIC for the Air Force's largest maintenance squadron supporting 15K personnel across three continents

So where do you find such a great first line? Like finding information for bullet statements, this too takes a bit of investigative work. You should consider what makes your unit unique among similar units on base, across the MAJCOM, or throughout the Air Force. Sometimes, your commander's biography will list the significance of your unit in the Air Force. You can also speak with your unit or base historian, search your installation's website, or even review Wikipedia. If you find something on Wikipedia, please verify the information with your historian.

Line 2—Office. Summarize the duties and responsibilities given by your supervisor. These are the direct tasks you are typically responsible for daily.

Line 3—Unit/Base. Summarize the services, products, and/or support you provide your unit or base.

Line 4—Base/External Agencies. Summarize the services, products, and/or support you provide your installation or outside organizations. Outside organizations can be a higher headquarters or agency in the local community. You can also use this line to include additional duties or deployments that extended over a significant portion of the evaluation period.

S I X
AF FORM 910 LINE-BY-LINE

Within this chapter, I will provide a line-by-line review of AF Form 910, *Enlisted Performance Report (AB thru TSgt)*. As you review the sections on the EPR, please ensure you first read the 'small print.' Each section identifies specific performance expectations, instructions, and criteria.

For Sections III-V, you will find more performance expectations than lines to write bullet statements. This allows you to be even more deliberate with the word picture (and perspective) you describe in each section and evaluation.

Remember to review your previous evaluations being considered for the Enlisted Force Distribution Panel and promotion board (as applicable) and AFH 36-2618 before writing your evaluation. If you accomplish similar tasks from the previous year, look to provide different perspectives or accomplishments to show a greater breadth of knowledge and experience. Also, review the expectations and responsibilities for the next higher grade and enlisted force structure tier in AFH 36-2618 to shape a deliberate word picture of your accomplishments. Provide a word picture that says you are ready to assume the next higher grade or opportunity in your career.

EPR Small Print

The EPR small print represents the performance expectations within a specific section in the evaluation. Like reading the small print on a contract, these areas provide the criteria for shaping the word picture in your evaluation.

Section III: Performance in Primary Duties/Training Requirements (Using AFH 36-2618, *The Enlisted Force Structure*, as the standard of performance expectations commensurate with the ratee's rank; assess to what degree the ratee complied with the following performance expectations.)

Task Knowledge/Proficiency—Consider the quality, results, and impact of the Airman's knowledge and ability to accomplish tasks.

Initiative/Motivation—Describes the degree of willingness to execute duties, motivate colleagues, and develop innovative new processes.

Skill Level Upgrade Training—Consider skill level awarding course, CDC timeliness completion, course exam results, and completion of core task training.

AF Form 910 Line-by-Line 53

Duty Position Requirements, Qualifications, and Certifications —Consider duty position qualifications, career field certifications (if applicable), and readiness requirements.

Training of Others—Consider the impact the Airman made training others.

Bullet Placement

Line 1—a strong bullet statement that captures an overview of your reporting period and connects with the duty description. If you do not provide an overview bullet statement, make sure this bullet statement is strong. This is often the first bullet statement a reviewer will see.

Lines 2–5—strong bullet statements that illustrate a deliberate word picture of duty performance. Show breadth of knowledge and experience throughout. Include specific accomplishments from the EPR small print, AFH 36-2618, and your duty description.

Line 6—a strong bullet statement that depicts (and summarizes) a different perspective of your duty performance.

EPR Small Print—Section IV: Followership/Leadership

Resource Utilization (e.g., Time Management, Equipment, Manpower, and Budget)—Consider how effectively the Airman utilizes resources to accomplish the mission.

Complies with/Enforces Standards—Consider personal adherence and enforcement of fitness standards, dress and personal appearance, customs and courtesies, and professional conduct.

Communication Skills—Describes how well the Airman receives and relays information, thoughts, and ideas up and down the chain of command (includes listening, reading, speaking, and writing skills); fosters an environment for open dialogue.

Caring, Respectful, and Dignified Environment (Teamwork)—Rate how well the Airman selflessly considers others, values diversity, and sets the stage for an environment of dignity and respect; to include promoting a healthy organizational climate.

Bullet Placement

Line 1—a varied but complementary bullet statement that captures a unique perspective of followership and/or leadership.

Line 2—a strong bullet statement that encompasses followership and/or leadership.

EPR Small Print—Section V: Whole Airman Concept

Air Force Core Values—Consider how well the Airman adopts, internalizes, and demonstrates our Air Force Core

Values of Integrity First, Service Before Self, and Excellence in All We Do.

Personal and Professional Development—Consider the amount of effort the Airman devoted to improving themselves and their work center/unit through education and involvement.

Esprit de corps and Community Relations—Consider how well the Airman promotes camaraderie, embraces esprit de corps, and acts as an Air Force ambassador.

Bullet Placement

Line 1—a varied but complementary bullet statement that captures a unique perspective of the whole Airman concept.

Line 2—a strong bullet statement that encompasses the whole Airman concept. Because this is the last bullet statement on the front page, you can also provide a strong descriptive bullet statement.

Additional Rater, Unit Commander, and Future Roles

Place your most significant achievements for the reporting period with the highest endorser/evaluator. You want your senior raters to provide the most impactful comments and bullet statements.

If you received multiple awards in a reporting period, you could display them in two ways. You can combine them all in one bullet statement in the last section, or you can place each award in separate sections with the most impactful (or higher level) award with the unit commander.

Section VIII, Line 1—second or third strongest bullet statement for the entire reporting period.

Section VIII, Line 2—first or second strongest bullet statement describing duty performance, leadership, mentoring, and/or an award win.

Section IX, Line 1—use your top performance bullet statement for the year, a strong descriptive or duty bullet statement, and/or an award win.

Descriptive Bullets

There should be at least one descriptive bullet statement in an evaluation. Because not every Airman will receive a Promote Now or Must Promote allocation, descriptive bullet statements highlight your competence, leadership, mentorship, judgment, etc. Align descriptions with responsibilities outlined for the next higher grade and/or enlisted force structure tier in AFH 36-2618. But be very cautious about your word choice. Descriptive bullet statements should complement and complete the assessment of your performance.

Future Roles

Future roles and assignments should best serve the Air Force and continue your development.[19] Do not haphazardly add roles and assignments to fill the blocks. Remember, your evaluation is one document that should provide deliberate information for future promotion potential and development.

If you are unsure which future roles might be available or applicable, first review the career pyramid in your CFETP and/or AFH 36-2618. Also, speak to your local or higher level functional manager and chief enlisted manager/squadron superintendent. They might even have listings of additional roles and assignments.

Final Thought

Never leave a line blank! Fill every line possible (unless purposefully leaving blank lines). Each line is an opportunity to describe your performance, capabilities, and potential to receive a specific career opportunity or assume the next higher grade.

Airmen who have evaluators with dual or multiple roles do not have to enter "THIS SECTION NOT USED" in the applicable sections. Additional comments can be added. Each section of the evaluation is considered separately.[20]

---

[19] Air Force Instruction 36-2406, Officer and Enlisted Evaluation Systems. Table 4.2., Item 29. United States Air Force. 8 November 2016. http://static.e-publishing.af.mil/production/1/af_a1/publication/afi36-2406/afi36-2406.pdf.

[20] Ibid., 4.13.6.

SEVEN
AF FORM 911 LINE-BY-LINE

Within this chapter, I will provide a line-by-line review of AF Form 911, *Enlisted Performance Report (MSgt thru SMSgt)*. (Similar information from the previous chapter might be repeated.) As you review the sections on the EPR, please ensure you first read the 'small print.' Each section identifies specific performance expectations, instructions, and criteria.

For Sections III-IV, you will find more performance expectations than lines to write bullet statements. This allows you to be even more deliberate with the word picture (and perspective) you describe in each section and evaluation.

Remember to review your previous evaluations being considered for your promotion board and AFH 36-2618 before writing your evaluation. If you accomplish similar tasks from the previous year, look to provide different perspectives or accomplishments to show a greater breadth of knowledge and experience. Also, review the expectations and responsibilities for the next higher grade in AFH 36-2618 to shape a deliberate word picture of your accomplishments.

Provide a word picture that says you are ready to assume the next higher grade or opportunity in your career.

EPR Small Print

The EPR small print represents the performance expectations within a specific section in the evaluation. Like reading the small print on a contract, these areas provide the criteria for shaping the word picture in your evaluation.

Section III: Performance in Leadership/Primary Duties/Followership/Training (Using AFH 36-2618, *The Enlisted Force Structure*, as the standard of performance expectations commensurate with the ratee's rank; assess to what degree the ratee complied with the following performance expectations.)

Mission Accomplishment—Consider the Airman's ability to lead and produce time, high quality/quantity, mission-oriented results.

Resource Utilization (e.g., time, management, equipment, manpower, and budget)—Consider how effectively the Airman leads their team to utilize their resources to accomplish the mission.

Team Building—Consider the amount of innovation, initiative, and motivation displayed by the Airman and their subordinates (collaboration).

Mentoring—Consider how well the Airman knows their subordinates, accepts personal responsibility for them, and is accountable for their professional development.

Communication Skills—Describe how well the Airman communicates (includes listening, reading, reading, speaking, and writing skills) in various mediums, translates superior's direction into specific tasks and responsibilities, fosters an environment for open dialogue, and enhances communication skills of subordinates.

Complies with/Enforces Standards—Consider personal adherence and how the Airman fosters an environment where everyone enforces fitness standards, dress and personal appearance, customs and courtesies, and professional conduct.

Duty Environments—Rate how well the Airman establishes and maintains caring, respectful, and dignified environments while valuing diversity; to include promoting a healthy organizational climate.

Training—Describes how well the Airman and their team complies with upgrade, duty position, and certification requirements.

Bullet Placement

Line 1—a strong bullet statement that captures an overview of your reporting period and connects with the duty description. If you do not provide an overview bullet statement, make sure this bullet statement is strong. This is often the first bullet statement a reviewer will see.

Lines 2–7—strong bullet statements that illustrate a specific word picture of duty performance. Show breadth of knowledge and experience throughout. Include specific accomplishments from the EPR small print, AFH 36-2618, and your duty description.

Line 8—a strong bullet statement that depicts (and summarizes) a different perspective of your duty performance.

EPR Small Print—Section IV: Whole Airman Concept

Air Force Core Values—Consider how well the Airman adopts, internalizes, demonstrates, and insists on adherence of our Air Force Core Values of Integrity First, Service Before Self, and Excellence in All We Do.

Personal and Professional Development—Consider effort the Airman devoted to improve their subordinates, their work center/unit and themselves.

Esprit de corps and Community Relations—Consider how well the Airman promotes camaraderie, enhances esprit de corps, and develops Air Force ambassadors.

Bullet Placement

Line 1—a varied but complementary bullet statement that captures a unique perspective of the whole Airman concept.

Line 2—a strong bullet statement that encompasses the whole Airman concept. Because this is the last bullet statement on the front page, you can also provide a strong descriptive bullet statement.

Additional Rater, Unit Commander, Future Roles, and Final Evaluator

Place your most significant achievements for the reporting period with the highest endorser/evaluator. You want your senior raters to provide the most impactful comments and bullet statements.

If you received multiple awards in a reporting period, you could display them in two ways. You can combine them all in one bullet statement in the last section, or you can place each award in separate sections with the most impactful (or higher level) award with the final evaluator.

Section VII, Line 1—second or third strongest bullet statement for the entire reporting period.

Section VII, Line 2—describe strong duty performance, leadership, mentoring, and/or an award win.

Section VIII, Line 1—a strong bullet statement that captures your performance in the unit, a descriptive bullet statement, and/or an award win.

Descriptive Bullets

There should be at least one descriptive bullet statement in an evaluation. Because not every Airman will receive a senior rater stratification/endorsement, descriptive bullet statements highlight your competence, leadership, mentorship, judgment, etc. Align descriptions with responsibilities outlined for the next higher grade in AFH 36-2618. But be very cautious about your word choice. Descriptive bullet statements should complement and complete the assessment of your performance.

Future Roles

Future roles and assignments should best serve the Air Force and continue your development.[21] Do not haphazardly add roles and assignments to fill the blocks. Remember, your evaluation is one document that should provide deliberate information for future promotion potential and development.

If you are unsure which future roles might be available or applicable, first review the career pyramid in your CFETP and/or AFH 36-2618. Also, speak to your local or higher level functional manager and chief enlisted manager/squadron

---

[21] Ibid., Table 4.2., Item 29.

superintendent. They might even have listings of additional roles and assignments.

Section XI, Line 1—use a descriptive bullet statement, your top performance bullet statement for the year, stratification (if applicable), and/or an award win.

Final Thought

Never leave a line blank! Fill every line possible (unless purposefully leaving blank lines). Each line is an opportunity to describe your performance, capabilities, and potential to receive a specific career opportunity or assume the next higher grade.

Airmen who have evaluators with dual or multiple roles do not have to enter "THIS SECTION NOT USED" in the applicable sections. Additional comments can be added. Each section of the evaluation is considered separately.[22]

---

[22] Ibid., 4.13.6.

E I G H T
PROMOTION BOARD SCORING

MSgt Snuffy thinks he plateaued in his career. Though he had a great start with Sergeant Jones, he never received another supervisor and mentor quite like him. After receiving the same board score for the previous two promotion cycles, he thought it was time to reach out to his old supervisor.

Chief Jones was now one of Sergeant Snuffy's MAJCOM functional managers. Sergeant Snuffy reached out to the chief and asked for his help. He asked if Chief Jones could review his records and let him know what he was missing. Chief Jones asked Sergeant Snuffy to send over his records and give him a few days before responding with feedback. Following are three things Chief Jones identified in Sergeant Snuffy's records:

1. No breadth of knowledge or experience. Though Sergeant Snuffy changed duty sections and bases, several of his evaluations listed five similar accomplishments again and again...and again. His accomplishments lacked breadth of knowledge and experience within his work centers, functional community, and performance as an Airman.

2. Lack of impactful accomplishments. Throughout Sergeant Snuffy's evaluations, there was little to no detail, quantitative data, or impactful results. His results were described as:

'100% accountability'

'100% compliant'

'enriched lives'

'bolstered knowledge'

'guaranteed accountability'

If you consider the six questions for finding the result of an accomplishment, one of the first questions you might ask in each of the 'results' above is 'how?.'

If there was 100% program accountability, what was the program like before Sergeant Snuffy's arrival? There might be quantitative data to gather. The same with '100% compliant.'

How did Sergeant Snuffy 'enrich lives?'

How does someone 'bolster knowledge?' Review my previous example of the education bullet statement in chapter four.

3. SNCO Deficiencies. Sergeant Snuffy was not showing himself as an active, visible SNCO.[23] He was not describing how his Airmen were being developed into better followers, leaders, and supervisors. Additionally, there was little mentoring, professional or personal development, awards, or unit, base, or community involvement within his evaluations.

Chief Jones told Sergeant Snuffy that there was nothing he could do to change the record that was written, but he can change the narrative of the next evaluation. The two talked about building a solid career map. Chief Jones also provided a summary of the messages he shared with Sergeant Snuffy when he was an Airman. Chief Jones also asked to review his evaluation bullet statements before him submitting them to his supervisor.

---

[23] Air Force Handbook 36-2618, The Enlisted Force Structure. Paragraph 4.6.4. United States Air Force. 5 July 2018. http://static.e-publishing.af.mil/production/1/af_a1/publication/afh36-2618/afh36-2618.pdf.

CONCLUSION

I appreciate the time you have taken to invest in not only your career but your Airmen as well. You organize, train, and equip great Airmen who live out our Air Force Core Values and accomplish the mission daily. You serve your country proudly and give your best effort in her defense. You owe it to yourself and your Airmen to accurately document your accomplishments. And I hope *Stripes: An Airman's Bullet Writing (and Career Management) Guide* will help you document your duty performance, meet your expectations and goals, and reach your Point B.

CONCLUSION

ACTION VERB LIST

On the following pages is an action verb list of over two thousand verbs. This verb list has been a part of my professional development toolbox since 2006. I hope it will help you as it has helped me.

accelerated	expressed	hastened	expedited
accepted	acknowledged	recognized	acclaimed
accomplished	consummated	completed	achieved
accounted	explained	elucidated	described
achieved	completed	attained	accomplished
acknowledged	admitted	recognized	affirmed
acquired	contracted	ordered	procured
acted	executed	officiated	transacted
activated	utilized	applied	exercised
actuated	propelled	drove	launched
adapted	arranged	allocated	adopted
added	adjoined	annexed	appended
addressed	applied	directed	approached
adjusted	modified	altered	revised
administered	discharged	contributed	dispensed
adopted	embraced	imitated	espoused
advanced	furthered	served	proceeded
advertised	contracted	engaged	reserved

advised	informed	considered	enlightened
advocated	recommended	sanctioned	approved
aided	assisted	benefited	helped
aligned	arranged	regulated	adjusted
allocated	adapted	appointed	designated
altered	modified	adjusted	revised
analyzed	examined	scrutinized	investigated
answered	acknowledged	confirmed	responded
anticipated	expected	prepared	predicted
applied	utilized	activated	exercised
appointed	chose	designated	nominated
appraised	informed	advised	notified
approved	authorized	certified	allowed
arbitrated	mediated	decided	reconciled
argued	debated	discussed	pleaded
arranged	assorted	grouped	indexed
articulated	voiced	expressed	pronounced
ascertained	demonstrated	proved	established
assembled	constructed	fabricated	created
assessed	negotiated	funded	deferred
assigned	constrained	compelled	absorbed
assisted	controlled	supervised	conducted
assured	guaranteed	affirmed	pledged
attained	completed	finished	concluded
attended	accompanied	succeeded	supported
audited	approved	balanced	validated
augmented	increased	enlarged	developed
authorized	allowed	sanctioned	empowered

Action Verb List

automated	programmed	streamlined	computerized
awarded	honored	granted	bestowed
balanced	equalized	offset	evened
beat	conquered	defeated	surpassed
began	initiated	started	embarked
bolstered	reinforced	supported	sustained
booked	contracted	engaged	reserved
boosted	advocated	validated	endorsed
bought	purchased	acquired	ordered
briefed	outlined	summarized	advised
broadened	increased	augmented	expanded
brought	accomplished	completed	executed
budgeted	purchased	acquired	ordered
built	constructed	fabricated	completed
calculated	planned	designed	steered
calibrated	regulated	tuned	balanced
catalogued	grouped	classified	assorted
categorized	assorted	classified	arranged
caught	seized	captured	snared
caused	produced	procured	effected
centralized	incorporated	assembled	gathered
chaired	administered	moderated	conducted
changed	altered	modified	adjusted
charged	assessed	expensed	supervised
charted	summarized	diagrammed	outlined
checked	curbed	limited	examined
chose	picked	named	elected
cited	named	mentioned	referred

clarified	simplified	facilitated	reduced
classified	grouped	catalogued	assorted
closed	ended	finished	concluded
coached	mentored	guided	trained
co-authored	designed	composed	initiated
co-founded	instituted	established	achieved
collaborated	cooperated	approved	assented
collated	accumulated	gathered	concentrated
collected	amassed	compiled	accumulated
combined	linked	joined	united
commanded	ordered	established	regulated
communicated	informed	announced	declared
compared	collated	compiled	measured
competed	contented	strived	opposed
compiled	grouped	catalogued	assorted
completed	concluded	built	achieved
complied	acceded	followed	conformed
composed	designed	authored	drafted
computed	sized	gauged	counted
conceived	instituted	achieved	founded
conceptualized	visualized	theorized	discerned
conciliated	soothed	appeased	mollified
concluded	ended	terminated	finished
condensed	consolidated	massed	accumulated
conditioned	changed	altered	modified
conducted	controlled	supervised	directed
conferred	granted	presented	bequeathed
confronted	opposed	disputed	challenged

Action Verb List

connected	combined	merged	united
conserved	preserved	maintained	sustained
considered	contemplated	investigated	weighed
consolidated	accumulated	massed	compiled
constructed	completed	assembled	created
consulted	discussed	conferred	counseled
contacted	communicated	reached	called
contributed	granted	conferred	bestowed
controlled	directed	guided	managed
conversed	spoke	discussed	talked
converted	transformed	remade	transfigured
convinced	converted	persuaded	indoctrinated
cooperated	united	banded	collaborated
coordinated	connected	balanced	offset
corrected	repaired	mended	fixed
correlated	activated	exercised	adjusted
corresponded	communicated	wrote	informed
counseled	recommended	advised	taught
counted	enumerated	reckoned	influenced
crafted	produced	assembled	constructed
created	constructed	crafted	completed
critiqued	examined	inquired	studied
cultivated	refined	finished	enlightened
customized	fitted	conformed	tailored
cut	penetrated	lanced	pierced
dealt	contracted	arranged	pledged
debated	argued	contested	discussed
decided	resolved	concluded	determined

decreased	reduced	contracted	diminished
deduced	concluded	postulated	understood
deferred	postponed	adjourned	prolonged
defined	outlined	fixed	limited
delegated	commissioned	authorized	appointed
delivered	transferred	rendered	issued
demonstrated	proved	established	ascertained
described	depicted	pictured	illustrated
designated	determined	destined	specified
designed	planned	aimed	calculated
detailed	specified	itemized	enumerated
detected	discovered	unearthed	encountered
determined	settled	decided	concluded
developed	perfected	advanced	refined
devised	invented	arranged	concocted
diagnosed	forecasted	predicted	prepared
diagrammed	sketched	pictured	described
differentiated	changed	altered	modified
digested	permeated	assimilated	metabolized
diminished	discontinued	abandoned	released
directed	controlled	supervised	conducted
disassembled	dismantled	divided	dismounted
disconnected	separated	detached	switched
discovered	encountered	detected	unearthed
discriminated	distinguished	sifted	separated
discussed	explained	considered	reviewed
dismantled	disconnected	divided	dismounted
dispatched	sent	transmitted	issued

dispensed	granted	conferred	bestowed
displayed	exhibited	exposed	demonstrated
disposed	adjusted	settled	adapted
disproved	refuted	revoked	invalidated
dissected	examined	operated	divided
disseminated	dissipated	scattered	dispersed
distinguished	characterized	identified	discerned
distributed	issued	scattered	arranged
diverted	deflected	distracted	redirected
documented	accounted	recorded	registered
drafted	made	designed	wrote
dramatized	performed	produced	portrayed
drew	attracted	enticed	pulled
drilled	rehearsed	taught	trained
dropped	discontinued	released	abandoned
drove	forced	propelled	pushed
duplicated	redoubled	multiplied	increased
earned	deserved	merited	warranted
edited	revised	corrected	composed
educated	trained	learned	accomplished
effected	concluded	completed	finalized
elected	picked	named	chose
electrified	delighted	impressed	excited
elevated	raised	lifted	improved
eliminated	dislodged	removed	extracted
employed	occupied	engaged	worked
enabled	sanctioned	enabled	confirmed
enacted	decided	determined	accomplished

encouraged	enlivened	motivated	energized
enforced	compelled	exerted	incited
engaged	contracted	occupied	absorbed
engineered	maneuvered	schemed	masterminded
enhanced	corrected	modified	updated
enjoyed	benefited	welcomed	relished
enlarged	expanded	augmented	developed
enlisted	secured	engaged	recruited
ensured	assured	guaranteed	secured
entered	admitted	installed	enrolled
entertained	engrossed	absorbed	engaged
equipped	implemented	readied	supplied
erected	produced	constructed	fabricated
established	fixed	secured	instituted
estimated	assessed	appraised	computed
evaluated	judged	appraised	estimated
examined	investigated	weighed	determined
exceeded	surpassed	beat	surmounted
executed	accomplished	completed	finished
exercised	utilized	applied	activated
exhibited	exposed	displayed	disclosed
expanded	increased	enlarged	augmented
expedited	hastened	accelerated	assisted
experienced	trained	matured	seasoned
experimented	examined	analyzed	probed
explained	elucidated	clarified	illustrated
explored	investigated	inquired	researched
expressed	declared	stated	asserted

Action Verb List 81

extracted	separated	transferred	eliminated
fabricated	constructed	completed	assembled
facilitated	simplified	reduced	clarified
familiarized	learned	enlightened	educated
fashioned	fabricated	made	manufactured
figured	computed	calculated	estimated
filed	applied	requested	submitted
filtered	refined	separated	purified
financed	provided	underwrote	arranged
fixed	attached	defined	assigned
followed up	ensured	proceeded	succeeded
forecasted	predicted	anticipated	foretold
foresaw	anticipated	planned	predicted
formed	shaped	molded	created
formulated	expressed	devised	drafted
forwarded	transferred	relocated	moved
fostered	prepared	nurtured	raised
found	established	created	discovered
founded	instituted	achieved	established
fulfilled	executed	completed	achieved
functioned as	performed	operated	officiated
gained	achieved	obtained	procured
gathered	congregated	assembled	collected
gave	bestowed	granted	donated
generated	made	created	originated
governed	superintended	managed	supervised
graded	rated	evaluated	judged
graphed	summarized	diagrammed	charted

greeted	saluted	welcomed	approached
grew	enlarged	expanded	increased
grossed	enlarged	increased	achieved
grouped	classified	assorted	arranged
guarded	protected	shielded	safeguarded
guided	controlled	supervised	conducted
handled	manipulated	rearranged	maneuvered
harmonized	coordinated	balanced	conformed
hastened	rushed	accelerated	dispatched
headed	determined	designated	destined
heightened	extended	elevated	enhanced
held	retained	defended	controlled
helped	assisted	upheld	sustained
highlighted	stressed	emphasized	pointed
hiked	explored	traveled	worked
hired	employed	engaged	procured
housed	built	established	harbored
hunted	pursued	searched	explored
identified	ascertained	analyzed	classified
illustrated	depicted	embellished	portrayed
imagined	hypothesized	conceived	pictured
immersed	submerged	engaged	intended
implemented	accomplished	fulfilled	achieved
improved	rectified	updated	corrected
improvised	constructed	created	produced
included	embodied	contained	encompassed
incorporated	instituted	achieved	founded
increased	extended	enhanced	heightened

Action Verb List

indexed	sorted	tabulated	catalogued
indicated	suggested	implied	insinuated
indoctrinated	propagandized	convinced	imbued
induced	affected	convinced	moved
inferred	deduced	implied	reasoned
influenced	affected	persuaded	convinced
informed	enlightened	educated	learned
initiated	inaugurated	enrolled	entered
injected	interjected	introduced	inserted
innovated	originated	devised	discovered
inspected	approved	validated	certified
inspired	galvanized	motivated	energized
installed	admitted	enrolled	entered
instigated	originated	launched	proposed
instituted	achieved	founded	chartered
insured	guaranteed	underwrote	safeguarded
integrated	connected	attached	joined
interfaced	meshed	weaved	twisted
interpreted	comprehended	examined	perceived
intervened	arbitrated	mediated	negotiated
interviewed	conversed	questioned	examined
introduced	acquainted	perceived	conducted
invented	originated	devised	discovered
inventoried	listed	catalogued	recorded
invested	enfolded	enclosed	infused
investigated	considered	weighed	determined
involved	included	entailed	comprised
isolated	separated	guaranteed	insulated

issued	published	distributed	dispatched
itemized	enumerated	recorded	arranged
joined	linked	connected	united
judged	tried	adjudicated	considered
kept	possessed	enjoyed	dominated
labeled	catalogued	assorted	grouped
labored	worked	strived	cultivated
launched	instituted	began	established
learned	experienced	discovered	mastered
lectured	addressed	expounded	taught
led	guided	ushered	conducted
leveled	equalized	balanced	compared
licensed	approved	authorized	accredited
listed	registered	certified	authorized
lobbied	influenced	induced	affected
located	discovered	detected	encountered
looked	scrutinized	expected	examined
lowered	reduced	decreased	diminished
made	fabricated	fashioned	manufactured
maintained	continued	extended	prolonged
managed	administered	directed	conducted
manipulated	maneuvered	managed	handled
mapped	summarized	diagrammed	charted
marked	inscribed	imprinted	labeled
marketed	sold	bargained	traded
mastered	conquered	subjugated	learned
matched	balanced	offset	evened
maximized	magnified	stressed	emphasized

Action Verb List

measured	gauged	calibrated	graded
mediated	arbitrated	reconciled	settled
mentioned	cited	named	specified
met	fulfilled	realized	achieved
minimized	lowered	decreased	diminished
mixed	combined	united	joined
mobilized	engaged	employed	selected
modeled	shaped	molded	created
moderated	reduced	curbed	modified
modified	changed	adjusted	altered
monitored	observed	watched	regulated
motivated	inspired	energized	encouraged
mounted	ascended	rose	scaled
moved	influenced	impressed	affected
multiplied	doubled	reproduced	increased
named	called	titled	labeled
navigated	guided	steered	conducted
negotiated	settled	bargained	moderated
netted	profited	accrued	gained
neutralized	offset	negated	nullified
noted	distinguished	celebrated	eminent
notified	apprised	informed	advised
observed	noticed	perceived	watched
obtained	completed	fulfilled	achieved
offered	proposed	extended	suggested
opened	initiated	began	inaugurated
operated	handled	conducted	produced
orchestrated	arranged	organized	managed

ordered	arranged	organized	placed
organized	established	instituted	planned
oriented	familiarized	informed	instructed
originated	established	launched	proposed
outlined	delineated	summarized	defined
overcame	conquered	captured	appropriated
overhauled	renovated	restored	rebuilt
oversaw	supervised	commanded	managed
paid	collected	gathered	accumulated
painted	portrayed	depicted	delineated
paraphrased	explained	interpreted	translated
participated	competed	contended	engaged
perceived	comprehended	understood	discerned
perfected	completed	developed	realized
performed	fulfilled	rendered	achieved
persuaded	converted	convinced	indoctrinated
photographed	illustrated	pictured	painted
piloted	steered	conducted	drove
pinpointed	determined	established	identified
pioneered	established	discovered	explored
placed	arranged	situated	established
planned	programmed	outlined	arranged
played	performed	competed	engaged
pointed	indicated	displayed	evinced
policed	patrolled	regulated	guarded
positioned	arranged	placed	established
practiced	experienced	versed	skilled
predicted	expected	anticipated	forecasted

Action Verb List

prepared	arranged	anticipated	contrived
prescribed	directed	appointed	guided
presented	granted	conferred	bestowed
preserved	guarded	secured	sheltered
presided	directed	controlled	officiated
prevailed	predominated	commanded	won
prevented	stopped	averted	blocked
prioritized	arranged	organized	ordered
probed	investigated	examined	explored
proceeded	progressed	advanced	initiated
processed	advanced	progressed	developed
procured	acquired	ordered	budgeted
produced	completed	assembled	created
profited	gained	benefited	recovered
programmed	computerized	automated	streamlined
prohibited	restricted	disallowed	prevented
projected	programmed	planned	outlined
promoted	recommended	approved	assisted
proofed	reviewed	analyzed	inspected
proofread	revised	corrected	rewrote
proposed	intended	designated	calculated
protected	defended	secured	safeguarded
proved	established	accepted	affirmed
provided	furnished	arranged	contributed
publicized	recognized	prominent	noted
published	distributed	printed	announced
purchased	acquired	ordered	procured
pushed	drove	inspired	propelled

qualified	experienced	capable	trained
questioned	inquired	sought	interrogated
quoted	cited	reviewed	mentioned
raised	constructed	prepared	elevated
ran	picked	elected	chosen
rated	appraised	evaluated	assessed
read	interpreted	translated	explained
realized	accomplished	achieved	performed
rearranged	organized	regulated	ordered
reasoned	argued	influenced	persuaded
reassembled	renewed	rallied	returned
recalled	eliminated	removed	expended
recapitulated	surrendered	submitted	yielded
received	obtained	secured	acquired
reckoned	calculated	computed	estimated
recognized	acclaimed	celebrated	accredited
recommended	endorsed	sanctioned	approved
reconciled	validated	balanced	audited
reconditioned	renovated	restored	rebuilt
reconstructed	reconditioned	renovated	restored
recorded	registered	chronicled	transcribed
recruited	secured	engaged	procured
redesigned	revised	amended	overhauled
reduced	diminished	decreased	curtailed
referred	designated	destined	headed
refined	cultivated	polished	accomplished
regrouped	recovered	triumphed	recaptured
regulated	arranged	ordered	classified

Action Verb List

rehabilitated	restored	stimulated	refreshed
reinforced	strengthened	intensified	bolstered
related	affiliated	associated	connected
relayed	informed	imparted	disclosed
remodeled	remade	improved	rebuilt
removed	transferred	eliminated	replaced
rendered	completed	finished	executed
renovated	remodeled	remade	improved
reordered	rearranged	reorganized	systematized
reorganized	arranged	organized	regulated
repaired	reconstructed	mended	corrected
repeated	restated	reiterated	recapitulated
rephrased	restated	reiterated	reworded
replaced	exchanged	substituted	transposed
replenished	inflated	expanded	boosted
reported	accounted	noted	recorded
represented	presented	denoted	portrayed
reproduced	duplicated	generated	multiplied
requested	inquired	petitioned	appealed
rescued	discovered	detected	unearthed
researched	investigated	examined	inquired
reshaped	formed	shaped	molded
resolved	determined	established	decided
responded	understood	retained	heeded
restated	repeated	reiterated	retold
restocked	refreshed	replenished	modernized
restored	reestablished	reinstated	reintroduced
restructured	rearranged	reorganized	overhauled

retold	recounted	narrated	reported
retrieved	discovered	detected	unearthed
returned	replaced	reinstated	reestablished
revamped	reworked	overhauled	redrafted
revealed	disclosed	exposed	unveiled
reviewed	analyzed	inspected	examined
revised	modified	adjusted	altered
revitalized	bolstered	energized	invigorated
revived	revitalized	refreshed	restored
reworked	adapted	modified	transformed
rewrote	restated	rephrased	reiterated
rotated	interchanged	alternated	substituted
routed	directed	detoured	bypassed
safeguarded	defended	protected	shielded
said	voiced	spoke	expressed
salvaged	retrieved	rescued	restored
saved	rescued	salvaged	liberated
saw	observed	witnessed	noticed
scanned	comprehended	interpreted	examined
scheduled	entered	registered	installed
scouted	explored	investigated	examined
screened	guarded	shielded	safeguarded
scrutinized	examined	inspected	investigated
searched	explored	investigated	researched
secured	safeguarded	defended	protected
selected	picked	named	elected
sent	bestowed	imparted	conferred
served	enlisted	enrolled	provided

Action Verb List 91

serviced	maintained	prepared	sustained
set	determined	decided	concluded
set up	prepared	contrived	arranged
shaped	formed	created	carved
sharpened	whetted	updated	amended
shipped	consigned	loaded	transmitted
shortened	condensed	abridged	summarized
showed	produced	presented	exhibited
sifted	filtered	screened	refined
signaled	gestured	indicated	suggested
signified	communicated	indicated	expressed
simplified	reduced	clarified	interpreted
sketched	planned	outlined	drafted
smoothed	evened	equalized	balanced
sold	purchased	finalized	transacted
solicited	canvassed	lobbied	influenced
solved	answered	deciphered	interpreted
sorted	ordered	arranged	distributed
sought	pursued	followed	aspired
sparked	enkindled	stimulated	motivated
spearheaded	initiated	commenced	instituted
specified	designated	stipulated	requested
spoke	communicated	verbalized	articulated
staffed	employed	enlisted	engaged
standardized	regulated	normalized	organized
started	founded	initiated	established
stated	declared	pronounced	asserted
stimulated	accelerated	motivated	encouraged

stirred	excited	provoked	stimulated
stopped	terminated	hindered	impeded
stored	packaged	gathered	collected
stowed	stored	packed	consigned
straightened	adjusted	corrected	leveled
streamlined	programmed	automated	computerized
strengthened	reinforced	restored	bolstered
stressed	emphasized	accentuated	highlighted
structured	arranged	configured	organized
studied	investigated	practiced	learned
submitted	relinquished	volunteered	proposed
succeeded	replaced	supplanted	superseded
suggested	implied	indicated	insinuated
summarized	encapsulated	abridged	condensed
superintended	oversaw	controlled	directed
supervised	controlled	directed	guided
supplied	outfitted	equipped	provided
supported	maintained	upheld	sustained
surmounted	transcended	mastered	overcame
surveyed	detected	perceived	viewed
switched	exchanged	substituted	transposed
synchronized	adjusted	adapted	harmonized
synthesized	consolidated	blended	coalesced
systematized	organized	coordinated	rearranged
tabulated	systematized	catalogued	complied
tackled	undertook	endeavored	exerted
talked	communicated	conversed	discussed
tallied	counted	recorded	calculated

Action Verb List 93

targeted	marked	imprinted	registered
taught	educated	instructed	trained
tended	protected	guarded	safeguarded
tested	inspected	examined	weighed
tightened	compressed	condensed	curtailed
timed	gauged	regulated	measured
took	acquired	obtained	seized
totaled	calculated	estimated	computed
toured	journeyed	visited	traveled
traced	pursued	investigated	determined
tracked	followed	pursued	shadowed
traded	exchanged	transposed	substituted
trained	educated	initiated	indoctrinated
transcribed	reproduced	deciphered	duplicated
transferred	conveyed	yielded	assigned
transformed	converted	transfigured	transmuted
translated	interpreted	reworded	paraphrased
transmitted	transported	transferred	conveyed
transported	conveyed	transmitted	transferred
traveled	seasoned	experienced	accomplished
treated	prescribed	ministered	attended
trimmed	pruned	sheared	truncated
troubleshot	explored	investigated	interrogated
turned	averted	diverted	transformed
tutored	educated	trained	enlightened
typed	transcribed	deciphered	prepared
uncovered	exposed	unveiled	unmasked
unearthed	discovered	detected	exposed

unfurled	uncoiled	expended	expanded
unified	affiliated	linked	connected
updated	corrected	amended	improved
underlined	underscored	emphasized	accentuated
upgraded	improved	modernized	modified
utilized	employed	manipulated	handled
validated	legalized	approved	certified
verbalized	articulated	vocalized	expressed
verified	authenticated	documented	corroborated
vitalized	animated	energized	invigorated
volunteered	offered	proposed	suggested
welcomed	appreciated	honored	esteemed
widened	broadened	expanded	increased
won	succeeded	triumphed	prevailed
worked	labored	operated	managed
wrote	composed	formulated	created

www.commissionpubs.com
info@commissionpubs.com

Made in the USA
Middletown, DE
20 December 2020